Standard Grade | Credit

Mathematics

Leckie × Leckie

First exam published in 2004.
Published by Leckie & Leckie Ltd, 3rd Floor, 4 Queen Street, Edinburgh EH2 1JE
tel: 0131 220 6831 fax: 0131 225 9987 enquiries@leckieandleckie.co.uk www.leckieandleckie.co.uk

ISBN 978-1-84372-639-5

A CIP Catalogue record for this book is available from the British Library.

Leckie & Leckie is a division of Huveaux plc.

Leckie & Leckie is grateful to the copyright holders, as credited at the back of the book, for permission to use their material. Every effort has been made to trace the copyright holders and to obtain their permission for the use of copyright material. Leckie & Leckie will gladly receive information enabling them to rectify any error or omission in subsequent editions.

[BLANK PAGE]

C

2500/405

NATIONAL QUALIFICATIONS 2004

FRIDAY, 7 MAY
1.30 PM – 2.25 PM

MATHEMATICS
STANDARD GRADE
Credit Level
Paper 1
(Non-calculator)

1 **You may NOT use a calculator**.

2 Answer as many questions as you can.

3 Full credit will be given only where the solution contains appropriate working.

4 Square-ruled paper is provided.

LIB 2500/405 6/33320

FORMULAE LIST

The roots of $ax^2 + bx + c = 0$ are $x = \dfrac{-b \pm \sqrt{(b^2 - 4ac)}}{2a}$

Sine rule: $\dfrac{a}{\sin A} = \dfrac{b}{\sin B} = \dfrac{c}{\sin C}$

Cosine rule: $a^2 = b^2 + c^2 - 2bc \cos A$ or $\cos A = \dfrac{b^2 + c^2 - a^2}{2bc}$

Area of a triangle: Area $= \frac{1}{2}ab \sin C$

Standard deviation: $s = \sqrt{\dfrac{\sum(x-\bar{x})^2}{n-1}} = \sqrt{\dfrac{\sum x^2 - (\sum x)^2/n}{n-1}}$, where n is the sample size.

KU	RE

1. Evaluate

$$6{\cdot}2 - (4{\cdot}53 - 1{\cdot}1).$$

2

2. Evaluate $\frac{2}{5}$ of $3\frac{1}{2} + \frac{4}{5}$

3

3. $A = 2x^2 - y^2$.

Calculate the value of A when $x = 3$ and $y = -4$.

2

4. Simplify $\frac{3}{m} + \frac{4}{(m+1)}$

3

5. The average monthly temperature in a holiday resort was recorded in degrees Celsius (°C).

Month	Jan	Feb	Mar	Apr	May	June	July	Aug	Sept	Oct	Nov	Dec
Average Temperature (°C)	1	8	8	10	15	22	23	24	20	14	9	4

Draw a suitable statistical diagram to illustrate the median and the quartiles of this data.

4

[Turn over

KU	RE

6. Marmalade is on special offer.

Each jar on special offer contains 12·5% more than the standard jar.

A jar on special offer contains 450 g of marmalade.

How much does the standard jar contain? **3** (KU)

7. John's school sells 1200 tickets for a raffle.

John buys 15 tickets.

John's church sells 1800 tickets for a raffle.

John buys 20 tickets.

In which raffle has he a better chance of winning the first prize? **3** (RE)

Show clearly all your working.

KU	RE

8. 7, -2, 5, 3, 8

In the sequence above, each term after the first two terms is the sum of the previous two terms.

For example: 3rd term = 5 = $7 + (-2)$

(*a*) A sequence follows the above rule.

The first term is x and the second term is y.

The fifth term is 5.

x, y, –, –, 5

Show that $2x + 3y = 5$ **2** (RE)

(*b*) Using the same x and y, another sequence follows the above rule.

The first term is y and the second term is x.

The sixth term is 17.

y, x, –, –, –, 17.

Write down another equation in x and y. **2** (RE)

(*c*) Find the values of x and y. **3** (RE)

9. The graph of $y = a \cos bx°$, $0 \leq x \leq 90$, is shown below.

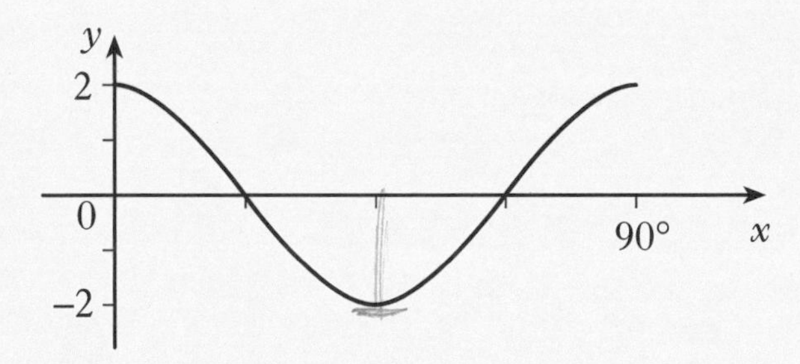

Write down the values of a and b. **2** (KU)

[Turn over for Questions 10, 11 and 12 on *Page six*

	KU	RE
10. Two variables x and y are connected by the relationship $y = ax + b$. Sketch a possible graph of y against x to illustrate this relationship when a and b are each less than zero.		3
11. (a) Simplify $2\sqrt{75}$	2	
(b) Evaluate $2^0 + 3^{-1}$.	2	
12. A piece of gold wire 10 centimetres long is made into a circle.		

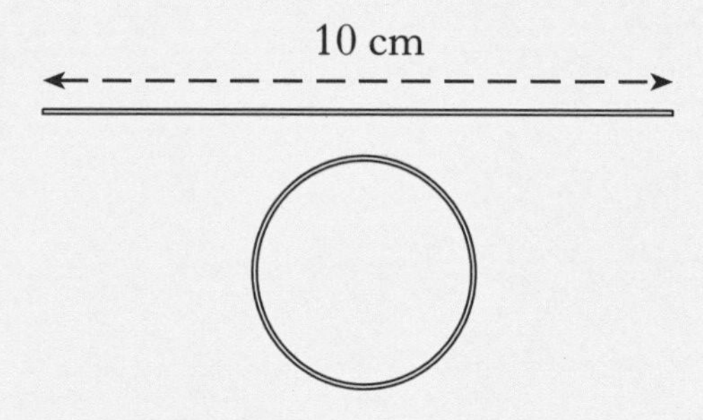

The circumference of the circle is equal to the length of the wire.

Show that the area of the circle is **exactly** $\frac{25}{\pi}$ square centimetres. (RE: 4)

[*END OF QUESTION PAPER*]

C

2500/406

NATIONAL QUALIFICATIONS 2004

FRIDAY, 7 MAY
2.45 PM – 4.05 PM

MATHEMATICS STANDARD GRADE

Credit Level
Paper 2

1 **You may use a calculator**.

2 Answer as many questions as you can.

3 Full credit will be given only where the solution contains appropriate working.

4 Square-ruled paper is provided.

LIB 2500/406 6/33320

FORMULAE LIST

The roots of $ax^2 + bx + c = 0$ are $x = \dfrac{-b \pm \sqrt{(b^2 - 4ac)}}{2a}$

Sine rule: $\dfrac{a}{\sin A} = \dfrac{b}{\sin B} = \dfrac{c}{\sin C}$

Cosine rule: $a^2 = b^2 + c^2 - 2bc \cos A$ or $\cos A = \dfrac{b^2 + c^2 - a^2}{2bc}$

Area of a triangle: Area $= \frac{1}{2} ab \sin C$

Standard deviation: $s = \sqrt{\dfrac{\sum (x - \bar{x})^2}{n-1}} = \sqrt{\dfrac{\sum x^2 - (\sum x)^2 / n}{n-1}}$, where n is the sample size.

KU | RE

1. Radio signals travel at a speed of 3×10^8 metres per second.

A radio signal from Earth to a space probe takes 8 hours.

What is the distance from Earth to the probe?

Give your answer **in scientific notation**. 4

2. A tank which holds 100 litres of water has a leak.

After 150 minutes, there is no water left in the tank.

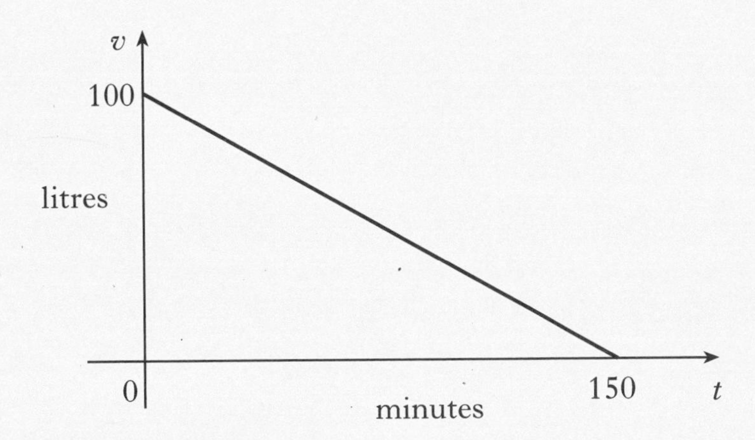

The above graph represents the volume of water (v litres) against time (t minutes).

(a) Find the equation of the line in terms of v and t. 3

(b) How many minutes does it take for the container to lose 30 litres of water? 3

3. Bottles of juice should contain 50 millilitres.

The contents of 7 bottles are checked in a random sample.

The actual volumes in millilitres are as shown below.

52, 50, 51, 49, 52, 53, 50

Calculate the mean and standard deviation of the sample. 4

KU	RE

4. 250 milligrams of a drug are given to a patient at 12 noon.

The amount of the drug in the bloodstream decreases by 20% every hour.

How many milligrams of the drug are in the bloodstream at 3pm? **3** (KU)

5. A helicopter, at point H, hovers between two aircraft carriers at points A and B which are 1500 metres apart.

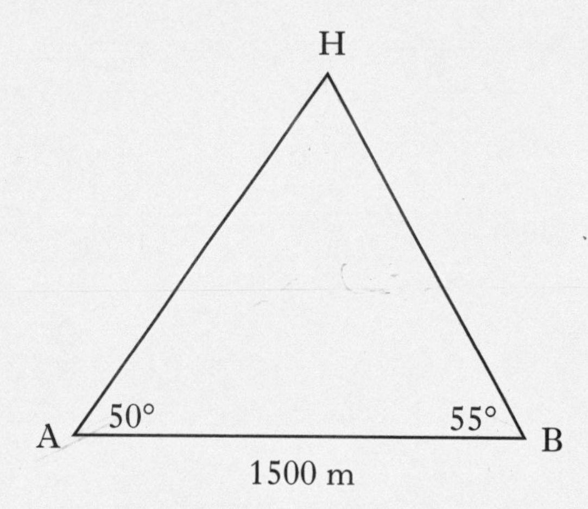

From carrier A, the angle of elevation of the helicopter is 50°.

From carrier B, the angle of elevation of the helicopter is 55°.

Calculate the distance from the helicopter to the nearer carrier. **4** (RE)

KU	RE

6. The diagram below shows a spotlight at point S, mounted 10 metres directly above a point P at the front edge of a stage.

The spotlight swings 45° from the vertical to illuminate another point Q, also at the front edge of the stage.

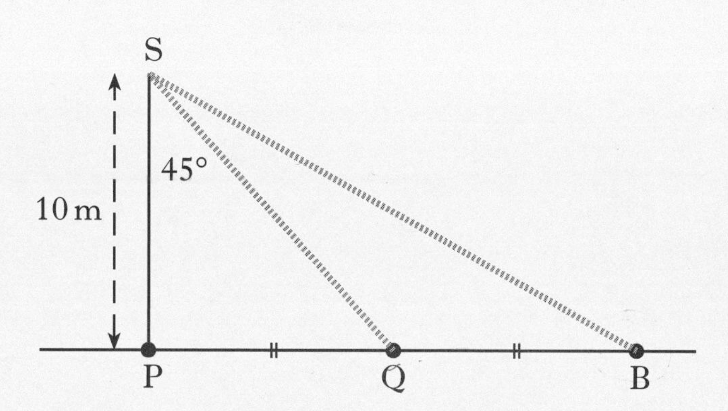

Through how many **more** degrees must the spotlight swing to illuminate a point B, where Q is the mid-point of PB? **5** (RE)

7. A square trapdoor of side 80 centimetres is held open by a rod as shown.

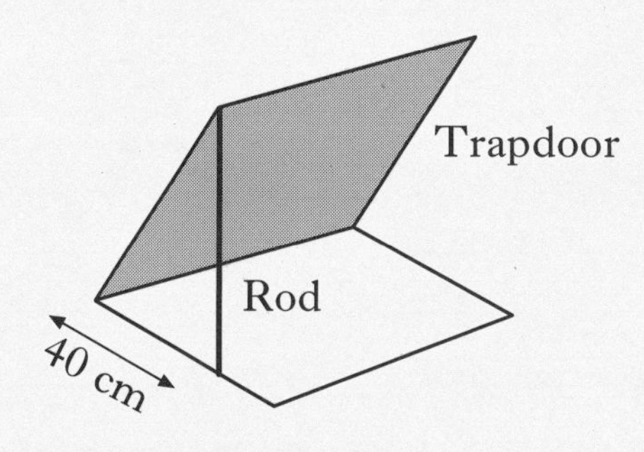

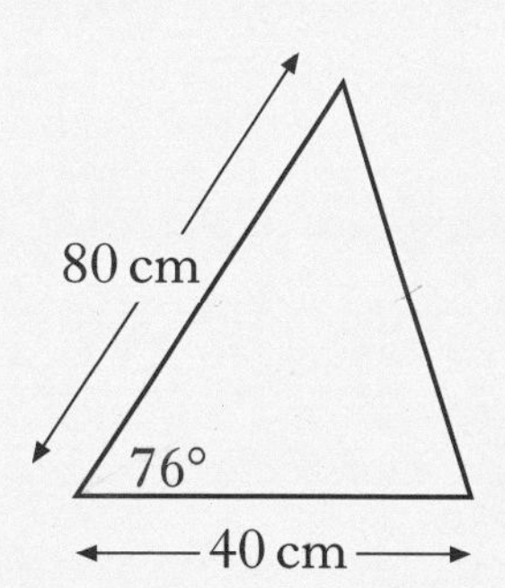

The rod is attached to a corner of the trapdoor and placed 40 centimetres along the edge of the opening.

The angle between the trapdoor and the opening is 76°.

Calculate the length of the rod to **2 significant figures**. **4** (KU)

[Turn over

KU	RE

8. The curved part of a doorway is an arc of a circle with radius 500 millimetres and centre C.

The height of the doorway to the top of the arc is 2000 millimetres.

The vertical edge of the doorway is 1800 millimetres.

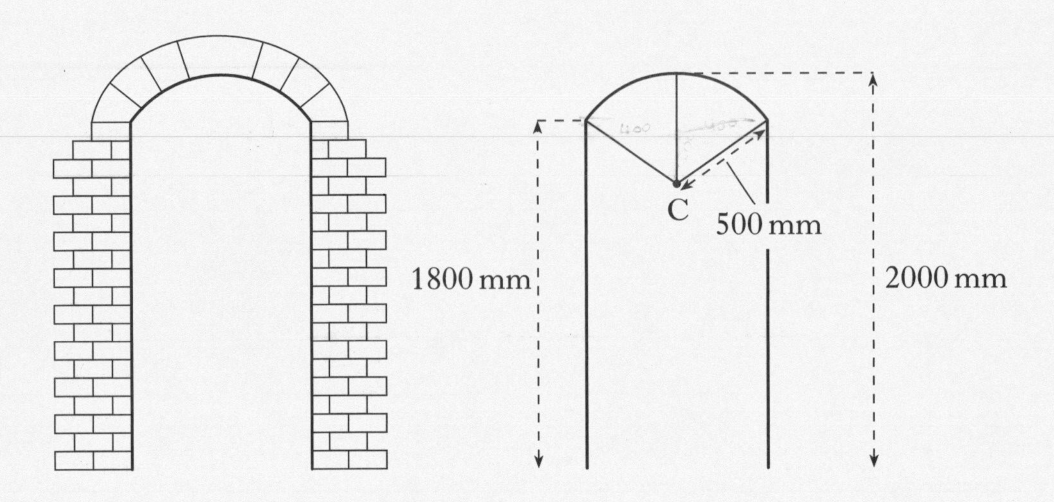

Calculate the width of the doorway. **5** (RE)

9. A gift box, 8 centimetres high, is prism shaped.

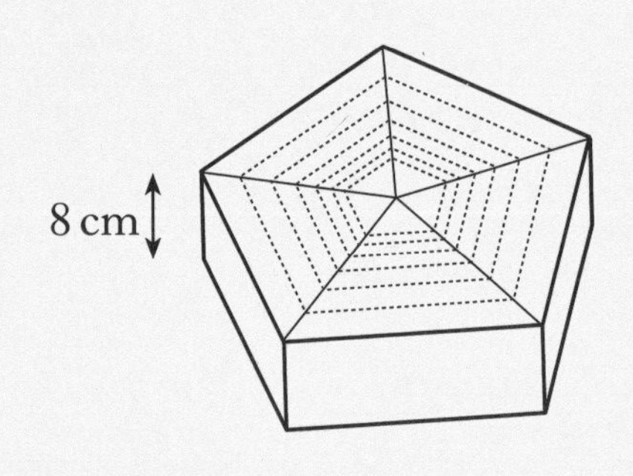

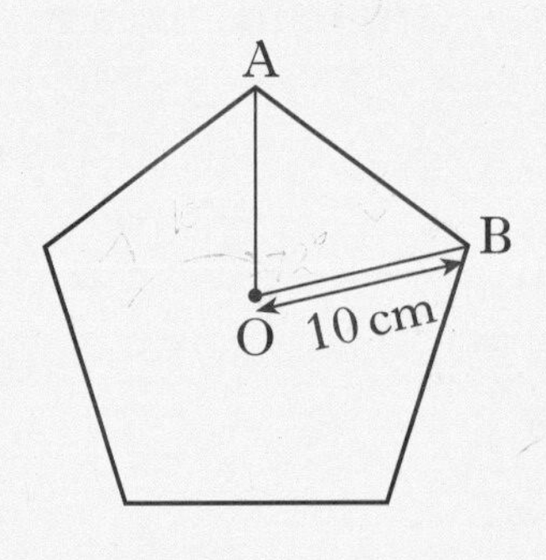

The uniform cross-section is a regular pentagon.

Each vertex of the pentagon is 10 centimetres from the centre O.

Calculate the volume of the box. **5** (KU)

	KU	RE
10. Solve **algebraically** the equation $$4\sin x° + 1 = -2 \qquad 0 \leq x < 360.$$	3	
11. A rectangular lawn has a path, 1 metre wide, on 3 sides as shown.		

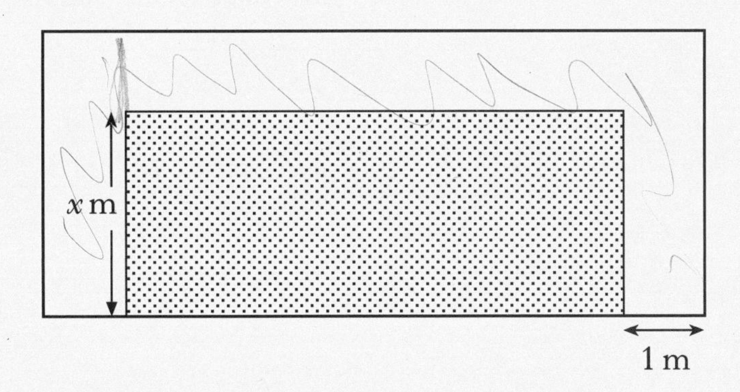

The breadth of the lawn is x metres.

The length of the lawn is three times its breadth.

The area of the lawn equals the area of the path.

	KU	RE
(*a*) Show that $3x^2 - 5x - 2 = 0$.		3
(*b*) Hence find the **length** of the lawn.		4

[*END OF QUESTION PAPER*]

[BLANK PAGE]

[BLANK PAGE]

C

2500/405

NATIONAL QUALIFICATIONS 2005

FRIDAY, 6 MAY
1.30 PM – 2.25 PM

MATHEMATICS STANDARD GRADE

Credit Level
Paper 1
(Non-calculator)

1 **You may NOT use a calculator**.

2 Answer as many questions as you can.

3 Full credit will be given only where the solution contains appropriate working.

4 Square-ruled paper is provided.

FORMULAE LIST

The roots of $ax^2 + bx + c = 0$ are $x = \dfrac{-b \pm \sqrt{(b^2 - 4ac)}}{2a}$

Sine rule: $\dfrac{a}{\sin A} = \dfrac{b}{\sin B} = \dfrac{c}{\sin C}$

Cosine rule: $a^2 = b^2 + c^2 - 2bc \cos A$ or $\cos A = \dfrac{b^2 + c^2 - a^2}{2bc}$

Area of a triangle: Area $= \frac{1}{2}ab \sin C$

Standard deviation: $s = \sqrt{\dfrac{\sum (x - \bar{x})^2}{n-1}} = \sqrt{\dfrac{\sum x^2 - (\sum x)^2 / n}{n-1}}$, where n is the sample size.

KU RE

1. Evaluate

$$3{\cdot}8 - (7{\cdot}36 \div 8).$$

2

2. Evaluate

$$2\tfrac{1}{3} + \tfrac{5}{6} \text{ of } 1\tfrac{2}{5}.$$

3

3. Evaluate

12·5% of £140.

2

4. Two identical dice are rolled simultaneously.

Find the probability that the total score on adding both numbers will be greater than 7 but less than 10.

2

[Turn over

	KU	RE

5. In an experiment involving two variables, the following values for x and y were recorded.

x	0	1	2	3	4
y	6	4	2	0	–2

The results were plotted, and a straight line was drawn through the points.

Find the gradient of the line **and** write down its equation. — KU 3

6. Solve the equation

$$\frac{2}{x} + 1 = 6.$$

— KU 3

7. The speeds (measured to the nearest 10 kilometres per hour) of 200 vehicles are recorded as shown.

Speed (km/hr)	30	40	50	60	70	80	90	100	110
Frequency	1	4	9	14	38	47	51	32	4

Construct a cumulative frequency table and hence find the median for this data. — KU 3

8. A number pattern is given below.

1st term: $2^2 - 0^2$
2nd term: $3^2 - 1^2$
3rd term: $4^2 - 2^2$

(*a*) Write down a similar expression for the 4th term. — RE 1

(*b*) Hence or otherwise find the nth term in its simplest form. — RE 3

KU	RE

9. (*a*) Emma puts £30 worth of petrol into the empty fuel tank of her car.

Petrol costs 75 pence per litre.
Her car uses 5 litres of petrol per hour, when she drives at a particular constant speed.

At this constant speed, how many litres of petrol will remain in the car after 3 hours? **2** (KU)

(*b*) The next week, Emma puts £20 worth of petrol into the empty fuel tank of her car.

Petrol costs c pence per litre.
Her car uses k litres of petrol per hour, when she drives at another constant speed.

Find a formula for R, the amount of petrol remaining in the car after t hours. **3** (RE)

10. A badge is made from a circle of radius 5 centimetres.

Segments are taken off the top and the bottom of the circle as shown.

The straight edges are parallel.

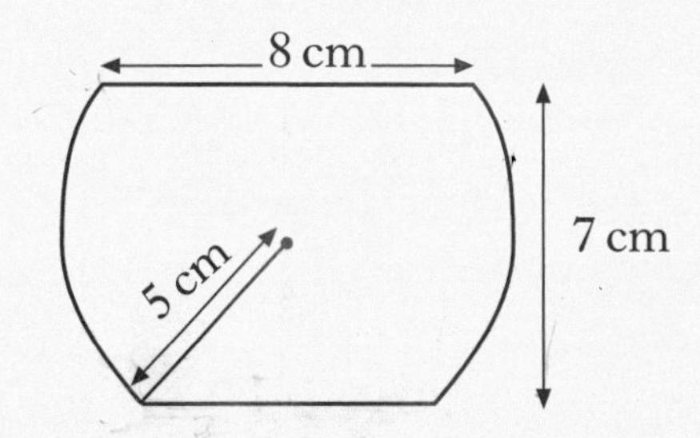

The badge measures 7 centimetres from the top to the bottom.
The top is 8 centimetres wide.

Calculate the width of the base. **5** (RE)

[Turn over

KU	RE

11. $f(x) = 4\sqrt{x} + \sqrt{2}$

(*a*) Find the value of $f(72)$ as a surd in its simplest form. **3** (KU)

(*b*) Find the value of t, given that $f(t) = 3\sqrt{2}$. **3** (RE)

12. The height of a triangle is $(2x - 5)$ centimetres and the base is $2x$ centimetres.

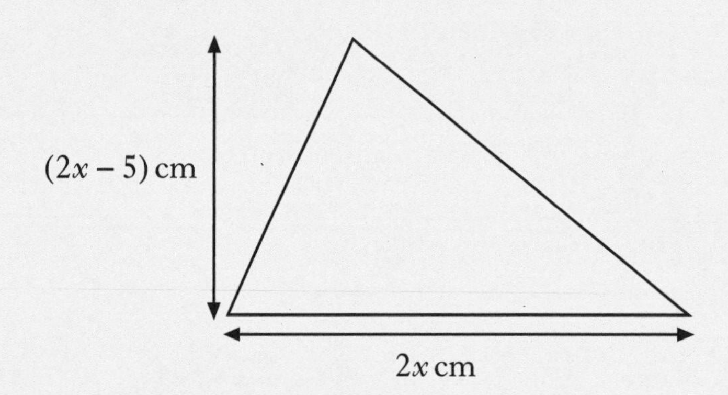

The area of the triangle is 7 square centimetres.

Calculate the value of x. **5** (RE)

[*END OF QUESTION PAPER*]

C

2500/406

NATIONAL QUALIFICATIONS 2005

FRIDAY, 6 MAY
2.45 PM – 4.05 PM

MATHEMATICS STANDARD GRADE

Credit Level
Paper 2

1 **You may use a calculator**.

2 Answer as many questions as you can.

3 Full credit will be given only where the solution contains appropriate working.

4 Square-ruled paper is provided.

LIB 2500/406 6/32970

FORMULAE LIST

The roots of $ax^2 + bx + c = 0$ are $x = \dfrac{-b \pm \sqrt{(b^2 - 4ac)}}{2a}$

Sine rule: $\dfrac{a}{\sin A} = \dfrac{b}{\sin B} = \dfrac{c}{\sin C}$

Cosine rule: $a^2 = b^2 + c^2 - 2bc \cos A$ or $\cos A = \dfrac{b^2 + c^2 - a^2}{2bc}$

Area of a triangle: Area $= \frac{1}{2}ab \sin C$

Standard deviation: $s = \sqrt{\dfrac{\sum(x - \bar{x})^2}{n-1}} = \sqrt{\dfrac{\sum x^2 - (\sum x)^2/n}{n-1}}$, where n is the sample size.

	KU	RE
1. $$E = mc^2.$$ Find the value of E when $m = 3{\cdot}6 \times 10^{-2}$ and $c = 3 \times 10^8$. Give your answer **in scientific notation**.	3	
2. The running times in minutes, of 6 television programmes are: 77 91 84 71 79 75. Calculate the mean and standard deviation of these times.	4	
3. Calculate the area of triangle PQR.	4	
4. Solve the equation $$x^2 + 2x = 9.$$ Give your answers **correct to 1 decimal place**.	3	

3.

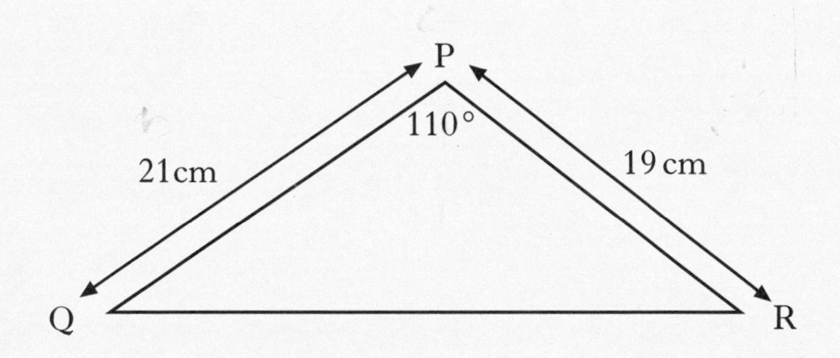

[Turn over

KU	RE

5. A triangular paving slab has measurements as shown.

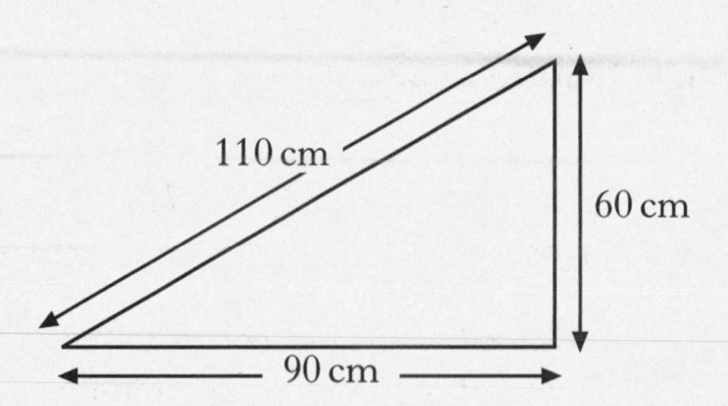

Is the slab in the shape of a right-angled triangle?

Show your working. 3

6. A vertical flagpole 12 metres high stands at the centre of the roof of a tower.

The tower is cuboid shaped with a square base of side 10 metres.

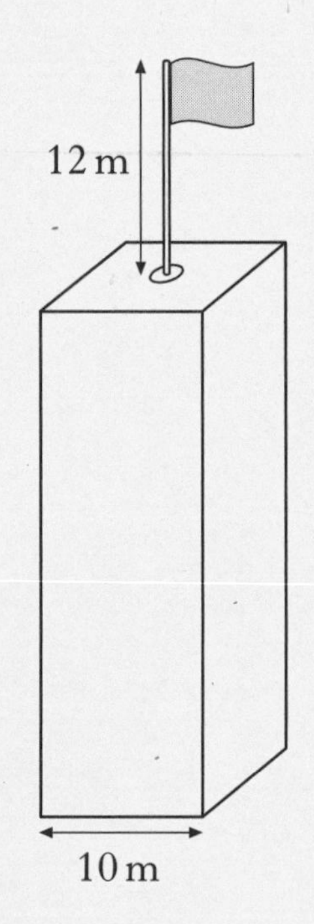

At a point P on the ground, 20 metres from the base of the tower, the top of the flagpole is just visible, as shown.

Calculate the height of the tower. 4

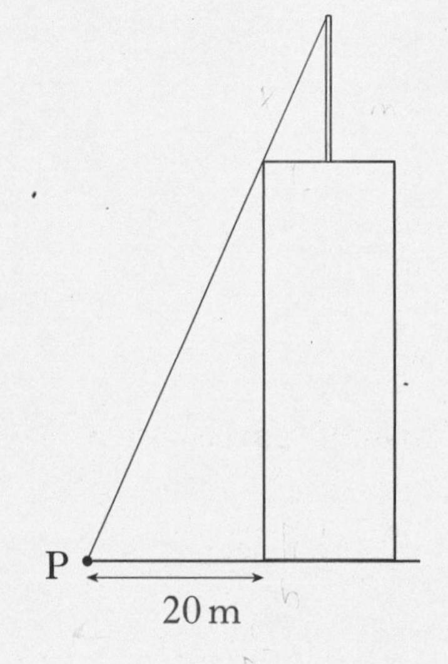

	KU	RE

7. David walks on a bearing of 050° from hostel A to a viewpoint V, 5 kilometres away.

Hostel B is due east of hostel A.

Susie walks on a bearing of 294° from hostel B to the same viewpoint.

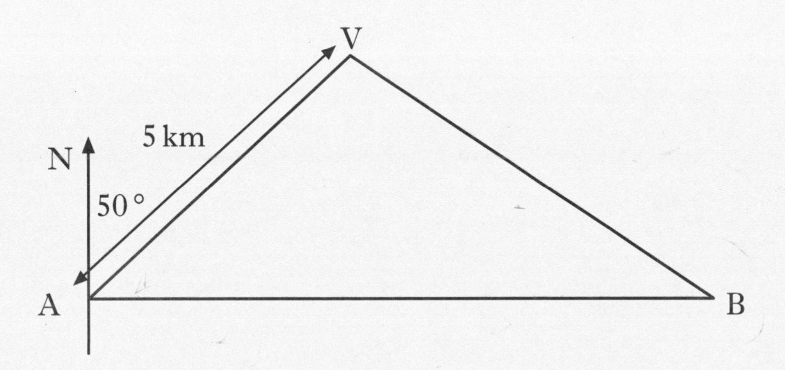

Calculate the length of AB, the distance between the two hostels. **5** (KU)

8. The side length of a cube is $2x$ centimetres.

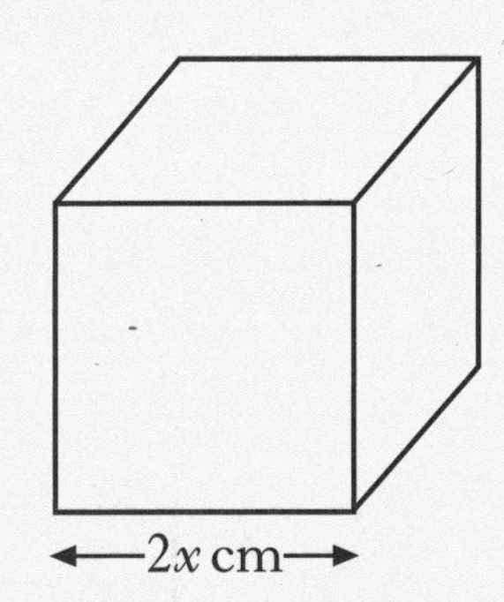

The expression for the volume in cubic centimetres is equal to the expression for the surface area in square centimetres.

Calculate the side length of the cube. **5** (RE)

[Turn over

	KU	RE

9. The monthly bill for a mobile phone is made up of a fixed rental plus call charges. Call charges vary as the time used.

The relationship between the monthly bill, y (pounds), and the time used, x (minutes) is represented in the graph below.

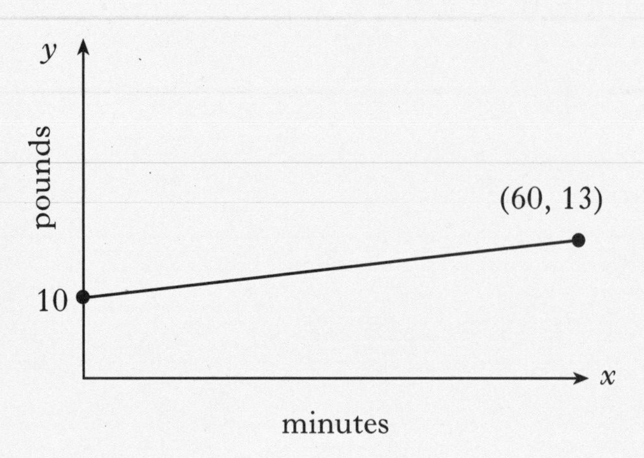

(*a*) Write down the fixed rental. **1**

(*b*) Find the call charge per minute. **3**

10. The chain of a demolition ball is 12·5 metres long.

When vertical, the end of the chain is 1·5 metres from the ground.

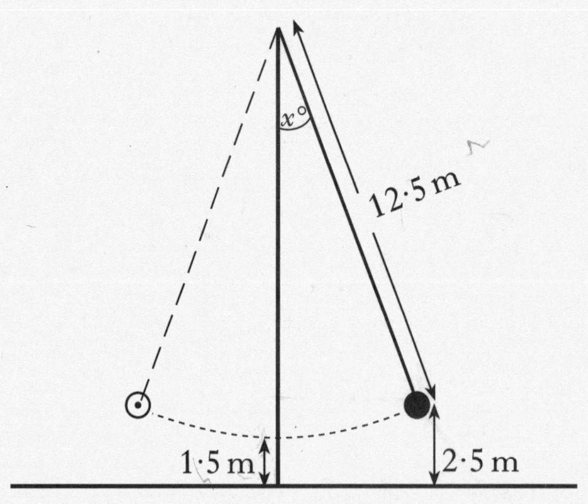

It swings to a maximum height of 2·5 metres above the ground on both sides.

(*a*) At this maximum height, show that the angle x°, which the chain makes with the vertical, is approximately 23°. **4**

(*b*) Calculate the maximum length of the arc through which the end of the chain swings. Give your answer **to 3 significant figures**. **4**

	KU	RE
11. (*a*) Solve algebraically the equation $\sqrt{3}\sin x° - 1 = 0 \qquad 0 \leq x < 360.$	3	
(*b*) Hence write down the solution of the equation $\sqrt{3}\sin 2x° - 1 = 0 \qquad 0 \leq x < 90.$		1

[*END OF QUESTION PAPER*]

[BLANK PAGE]

[BLANK PAGE]

C

2500/405

NATIONAL QUALIFICATIONS 2006

FRIDAY, 5 MAY
1.30 PM – 2.25 PM

MATHEMATICS STANDARD GRADE

Credit Level
Paper 1
(Non-calculator)

1 **You may NOT use a calculator**.

2 Answer as many questions as you can.

3 Full credit will be given only where the solution contains appropriate working.

4 Square-ruled paper is provided.

LI 2500/405 6/31570

FORMULAE LIST

The roots of $ax^2 + bx + c = 0$ are $x = \dfrac{-b \pm \sqrt{(b^2 - 4ac)}}{2a}$

Sine rule: $\dfrac{a}{\sin A} = \dfrac{b}{\sin B} = \dfrac{c}{\sin C}$

Cosine rule: $a^2 = b^2 + c^2 - 2bc \cos A$ or $\cos A = \dfrac{b^2 + c^2 - a^2}{2bc}$

Area of a triangle: Area $= \frac{1}{2}ab \sin C$

Standard deviation: $s = \sqrt{\dfrac{\sum(x - \bar{x})^2}{n-1}} = \sqrt{\dfrac{\sum x^2 - (\sum x)^2 / n}{n-1}}$, where n is the sample size.

KU	RE

1. Evaluate

$$56{\cdot}4 - 1{\cdot}25 \times 40.$$

2

2. Evaluate

$$1\tfrac{3}{5} + 2\tfrac{4}{7}.$$

2

3. Given that $f(x) = 4 - x^2$, evaluate $f(-3)$.

2

4.

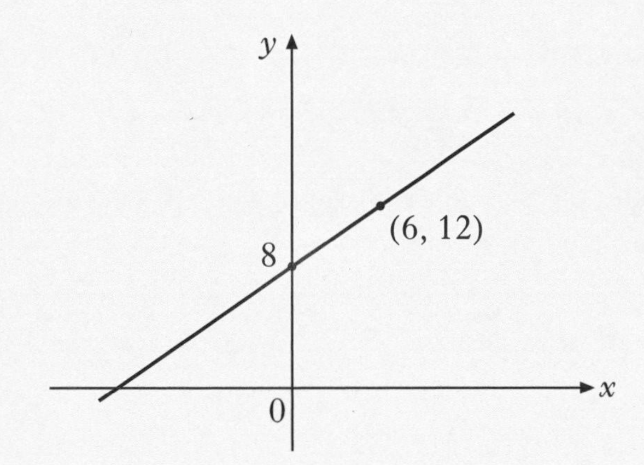

Find the equation of the given straight line.

3

[Turn over

	KU	RE

5. (*a*) Factorise

$$4x^2 - y^2.$$ **1**

(*b*) Hence simplify

$$\frac{4x^2 - y^2}{6x + 3y}.$$ **2**

6. Solve the equation

$$x - 2(x + 1) = 8.$$ **3**

7. Coffee is sold in regular cups and large cups.

The two cups are mathematically similar in shape.

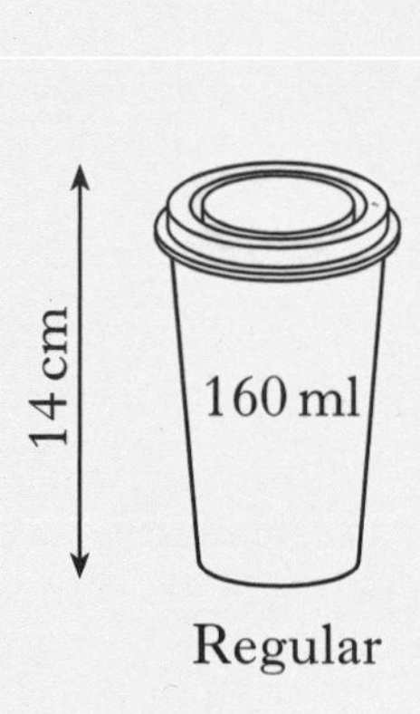

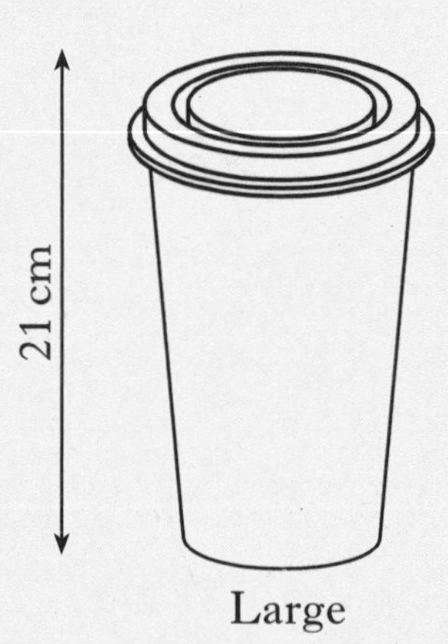

The regular cup is 14 centimetres high and holds 160 millilitres.

The large cup is 21 centimetres high.

Calculate how many millilitres the large cup holds. **4**

KU	RE

8. The graph of $y = x^2$ has been moved to the position shown in Figure 1.

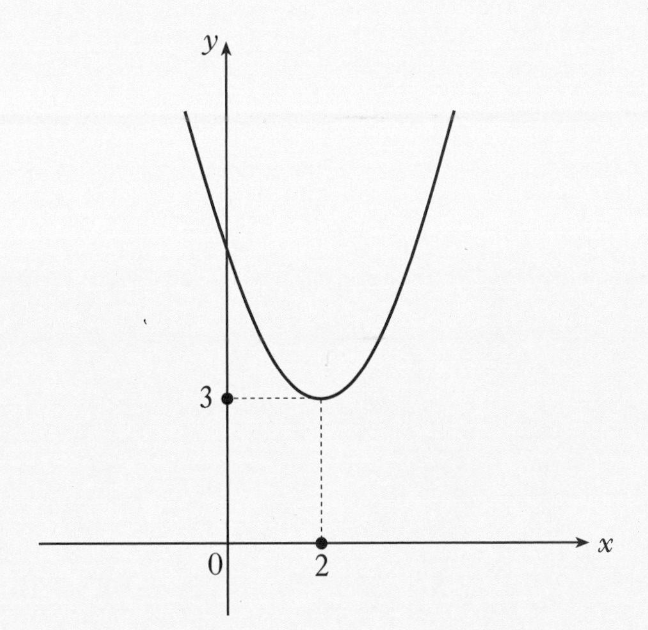

Figure 1

The equation of this graph is $y = (x - 2)^2 + 3$.

The graph of $y = x^2$ has now been moved to the position shown in Figure 2.

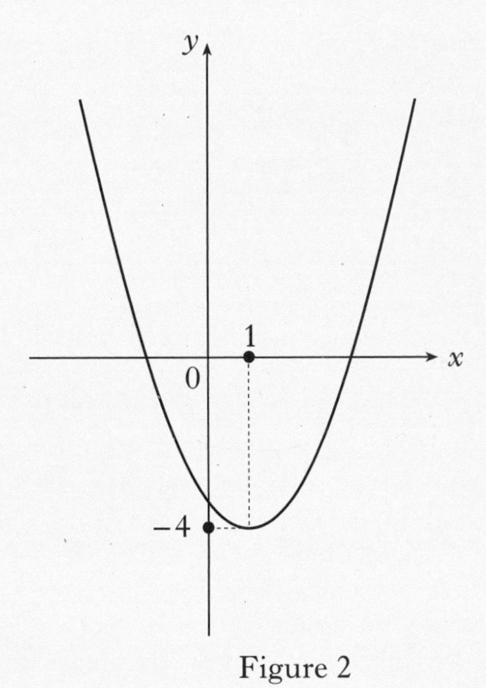

Figure 2

Write down the equation of the graph in Figure 2. **2**

[Turn over

	KU	RE
9. Euan plays in a snooker tournament which consists of 20 games.		
He wins x games and loses y games.		
(*a*) Write down an equation in x and y to illustrate this information.	1	
(*b*) He is paid £5 for each game he wins and £2 for each game he loses. He is paid a **total** of £79. Write down another equation in x and y to illustrate this information.		2
(*c*) How many games did Euan **win**?		3

10. Triangle ABC is right-angled at B.

The dimensions are as shown.

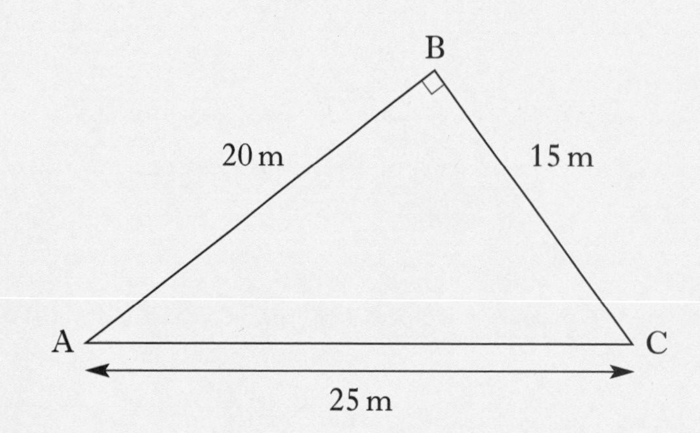

	KU	RE
(*a*) Calculate the area of triangle ABC.	1	

(*b*) BD, the height of triangle ABC, is drawn as shown.

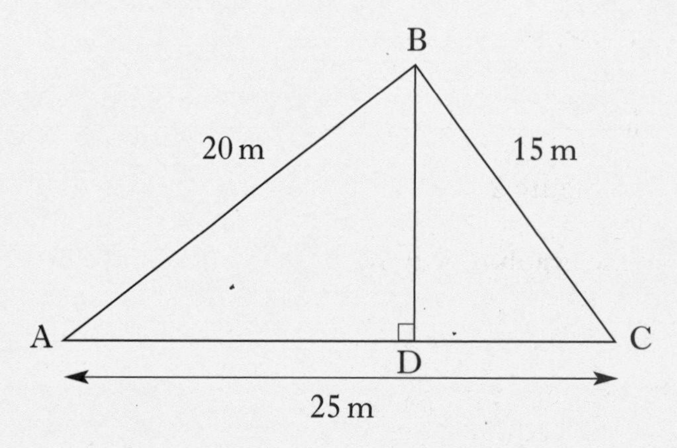

	KU	RE
Use your answer to part (*a*) to calculate the height BD.		3

KU	RE

11. (*a*) One session at the Leisure Centre costs £3.

£3 per session

Write down an algebraic expression for the cost of x sessions. — RE 1

(*b*) The Leisure Centre also offers a monthly card costing £20. The **first 6** sessions are then free, with each additional session costing £2.

Monthly card
£20

* first 6 sessions **free**
* each additional session £2

(i) Find the **total** cost of a monthly card and 15 sessions. — KU 1

(ii) Write down an algebraic expression for the **total** cost of a monthly card and x **sessions**, where x is greater than 6. — RE 2

(*c*) Find the minimum number of sessions required for the monthly card to be the cheaper option.

Show all working. — RE 3

[*END OF QUESTION PAPER*]

[BLANK PAGE]

C

2500/406

NATIONAL
QUALIFICATIONS
2006

FRIDAY, 5 MAY
2.45 PM – 4.05 PM

MATHEMATICS
STANDARD GRADE
Credit Level
Paper 2

1 **You may use a calculator**.

2 Answer as many questions as you can.

3 Full credit will be given only where the solution contains appropriate working.

4 Square-ruled paper is provided.

LI 2500/406 6/31570

FORMULAE LIST

The roots of $ax^2 + bx + c = 0$ are $x = \dfrac{-b \pm \sqrt{(b^2 - 4ac)}}{2a}$

Sine rule: $\dfrac{a}{\sin \text{A}} = \dfrac{b}{\sin \text{B}} = \dfrac{c}{\sin \text{C}}$

Cosine rule: $a^2 = b^2 + c^2 - 2bc \cos \text{A}$ or $\cos \text{A} = \dfrac{b^2 + c^2 - a^2}{2bc}$

Area of a triangle: Area $= \frac{1}{2}ab \sin \text{C}$

Standard deviation: $s = \sqrt{\dfrac{\sum (x - \bar{x})^2}{n-1}} = \sqrt{\dfrac{\sum x^2 - (\sum x)^2/n}{n-1}}$, where n is the sample size.

	KU	RE
1. The orbit of a planet around a star is circular. The radius of the orbit is $4 \cdot 96 \times 10^7$ kilometres. Calculate the circumference of the orbit. Give your answer **in scientific notation**.	3	
2. (*a*) The pulse rates, in beats per minute, of 6 adults in a hospital waiting area are: 68 73 86 72 82 78. Calculate the mean and standard deviation of this data.	3	
(*b*) 6 children in the same waiting area have a mean pulse rate of 89·6 beats per minute and a standard deviation of 5·4. Make **two** valid comparisons between the children's pulse rates and those of the adults.		2
3. Harry bids successfully for a painting at an auction. An "auction tax" of 8% is added to his bid price. He pays £324 in total. Calculate his bid price.	3	

[Turn over

KU	RE

4. (*a*) Expand and simplify

$$(x+4)(3x-1).$$

1

(*b*) Expand

$$m^{\frac{1}{2}}(2+m^2).$$

2

(*c*) Simplify, leaving your answer as a surd

$$2\sqrt{20}-3\sqrt{5}.$$

2

5. ST, a vertical pole 2 metres high, is situated at the corner of a rectangular garden, PQRS.

RS is 8 metres long and QR is 12 metres long.

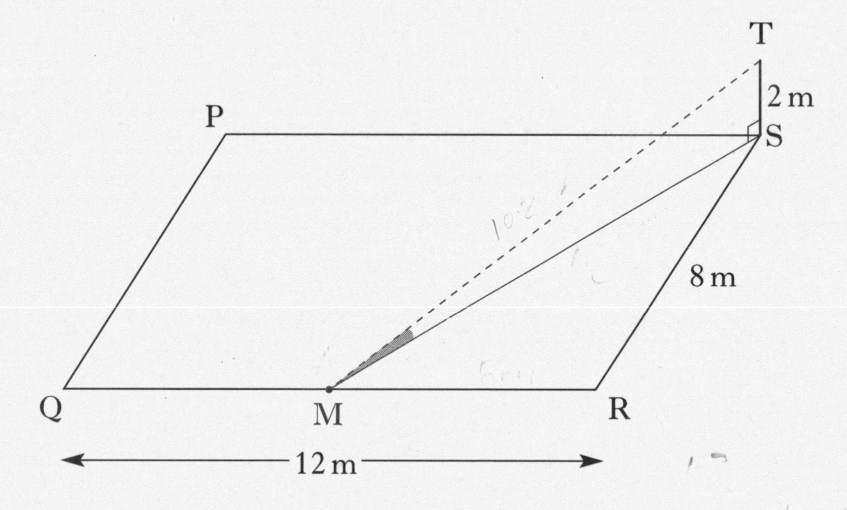

The pole casts a shadow over the garden.

The shadow reaches M, the midpoint of QR.

Calculate the size of the shaded angle TMS.

4

	KU	RE

6. (*a*) There are three mooring points A, B and C on Lake Sorling.

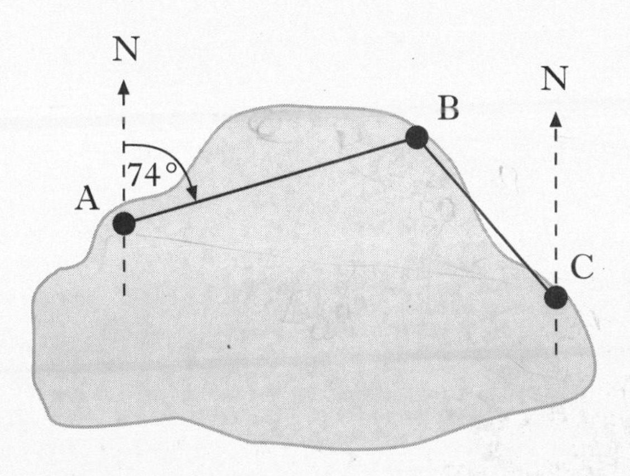

From A, the bearing of B is 074°.

From C, the bearing of B is 310°.

Calculate the size of angle ABC. **2** (RE)

(*b*) B is 230 metres from A and 110 metres from C.

Calculate the direct distance from A to C.

Give your answer **to 3 significant figures**. **4** (RE)

7. (*a*) A block of copper 18 centimetres long is prism shaped as shown.

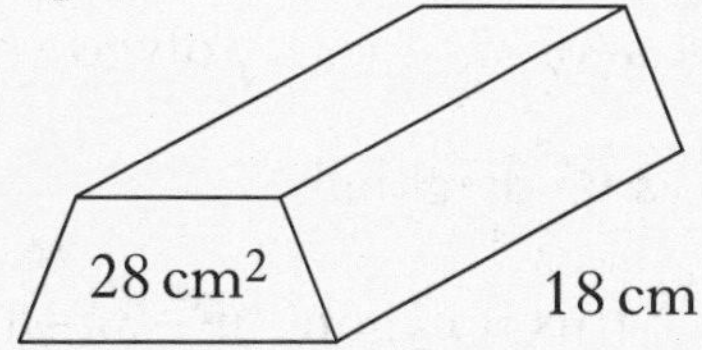

The area of its cross section is 28 square centimetres.

Find the volume of the block. **1** (KU)

(*b*) The block is melted down to make a cylindrical cable of diameter 14 **millimetres**.

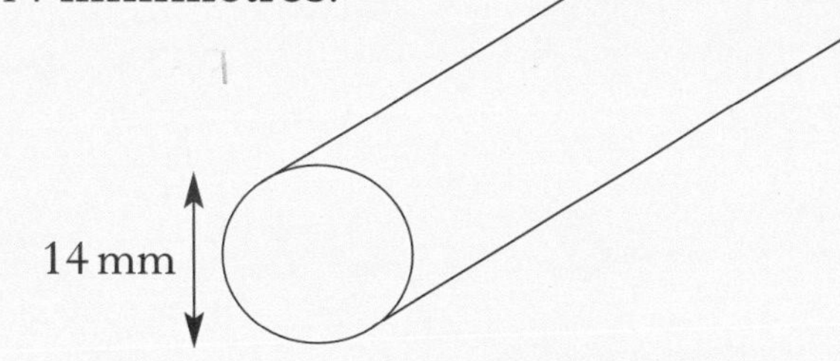

Calculate the length of the cable. **4** (RE)

KU	RE

8. A set of scales has a circular dial.

The pointer is 9 centimetres long.

The tip of the pointer moves through an arc of 2 centimetres for each 100 grams of weight on the scales.

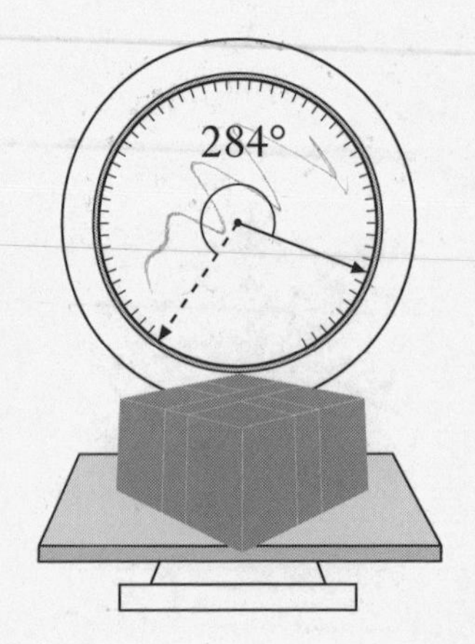

A parcel, placed on the scales, moves the pointer through an angle of 284°.

Calculate the weight of the parcel. **4** (RE)

9. The number of diagonals, d, in a polygon of n sides is given by the formula

$$d = \frac{1}{2}n(n-3).$$

(*a*) How many diagonals does a polygon of 7 sides have? **2** (KU)

(*b*) A polygon has 65 diagonals.

Show that for this polygon, $n^2 - 3n - 130 = 0$. **2** (RE)

(*c*) Hence find the number of sides in this polygon. **3** (RE)

	KU	RE

10. Emma goes on the "Big Eye".

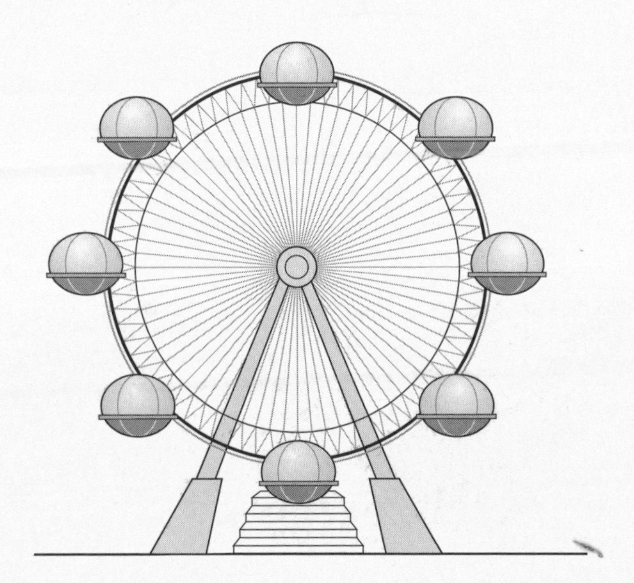

Her height, h metres, above the ground is given by the formula

$$h = -31 \cos t° + 33$$

where t is the number of seconds after the start.

(*a*) Calculate Emma's height above the ground 20 seconds after the start. — KU 2

(*b*) When will Emma first reach a height of 60 metres above the ground? — RE 3

(*c*) When will she next be at a height of 60 metres above the ground? — RE 1

[Turn over for Question 11 on *Page eight*

KU	RE

11. In triangle ABC,

BC = 8 centimetres,

AC = 6 centimetres and

PQ is parallel to BC.

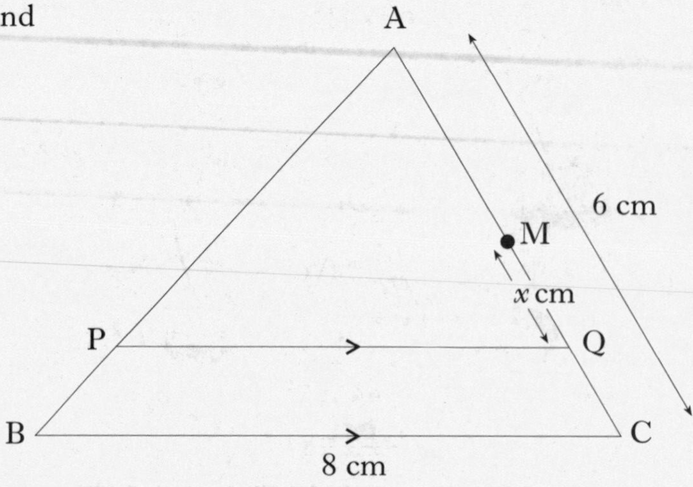

M is the midpoint of AC.

Q lies on AC, x centimetres from M, as shown on the diagram.

(*a*) Write down an expression for the length of AQ. **1**

(*b*) Show that PQ = $(4 + \frac{4}{3}x)$ centimetres. **3**

[*END OF QUESTION PAPER*]

[BLANK PAGE]

C

2500/405

NATIONAL
QUALIFICATIONS
2007

THURSDAY, 3 MAY
1.30 PM – 2.25 PM

MATHEMATICS
STANDARD GRADE
Credit Level
Paper 1
(Non-calculator)

1 **You may NOT use a calculator**.

2 Answer as many questions as you can.

3 Full credit will be given only where the solution contains appropriate working.

4 Square-ruled paper is provided.

LI 2500/405 6/31570

FORMULAE LIST

The roots of $ax^2 + bx + c = 0$ are $x = \dfrac{-b \pm \sqrt{(b^2 - 4ac)}}{2a}$

Sine rule: $\dfrac{a}{\text{Sin A}} = \dfrac{b}{\text{Sin B}} = \dfrac{c}{\text{Sin C}}$

Cosine rule: $a^2 = b^2 + c^2 - 2bc \cos \text{A}$ or $\cos \text{A} = \dfrac{b^2 + c^2 - a^2}{2bc}$

Area of a triangle: Area $= \frac{1}{2} ab \sin \text{C}$

Standard deviation: $s = \sqrt{\dfrac{\sum(x - \bar{x})^2}{n-1}} = \sqrt{\dfrac{\sum x^2 - (\sum x)^2/n}{n-1}}$, where n is the sample size.

	KU	RE
1. Evaluate $$6{\cdot}04 + 3{\cdot}72 \times 20.$$	2	
2. Evaluate $$3\tfrac{1}{6} \div 1\tfrac{2}{3}$$	2	
3. There are 400 people in a studio audience. The probability that a person chosen at random from this audience is male is $\frac{5}{8}$ How many males are in this audience?		2
4. $$P = \frac{2(m-4)}{3}$$ Change the subject of the formula to m.	3	
5. Remove brackets and simplify $$(2x+3)^2 - 3(x^2 - 6).$$	3	

[Turn over

	KU	RE

6. A taxi fare consists of a £2 "call-out" charge **plus** a fixed amount per kilometre.

The graph shows the fare, f pounds for a journey of d kilometres.

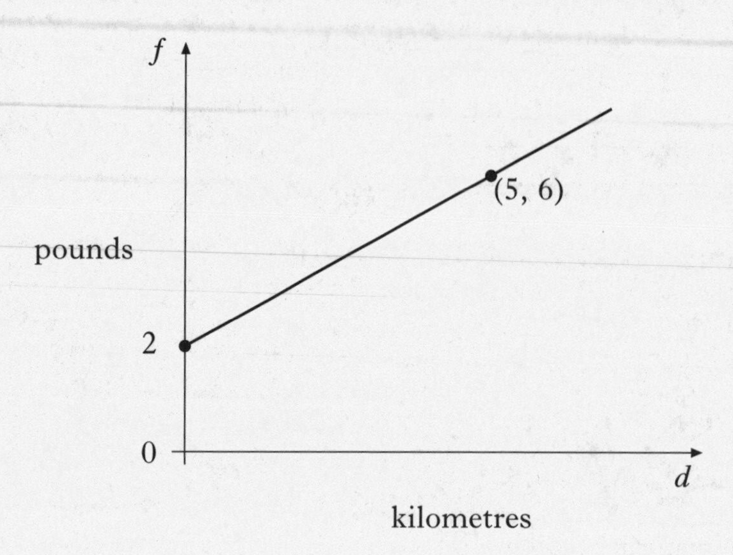

The taxi fare for a 5 kilometre journey is £6.

Find the equation of the straight line in terms of d and f. **4**

7. Remove brackets and simplify

$$a^{\frac{1}{2}}(a^{\frac{1}{2}} - 2).$$ **2**

	KU	RE

8. Mick needs an ironing board.

He sees one in a catalogue with measurements as shown in the diagram below.

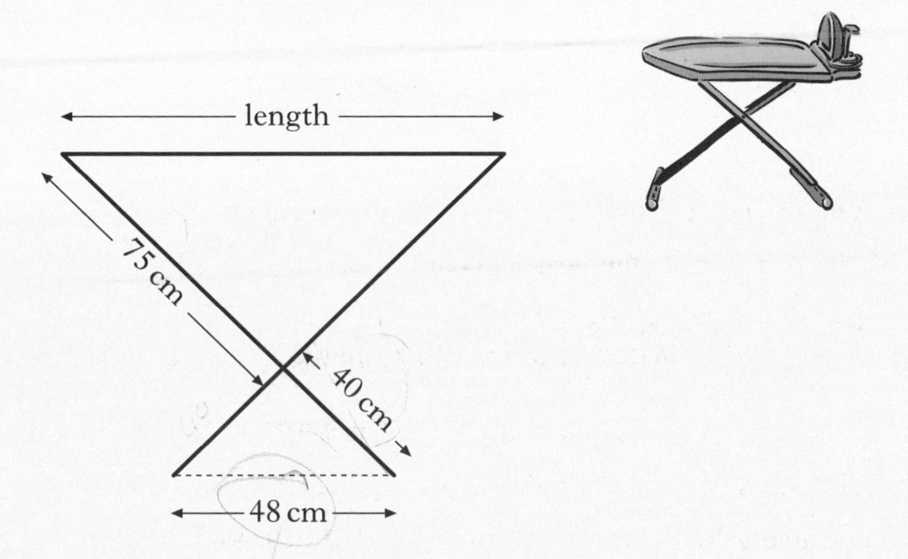

When the ironing board is set up, two similar triangles are formed.

Mick wants an ironing board which is at least 80 centimetres in length.

Does this ironing board meet Mick's requirements?

Show all your working. 3

9. A square of side x centimetres has a diagonal 6 centimetres long.

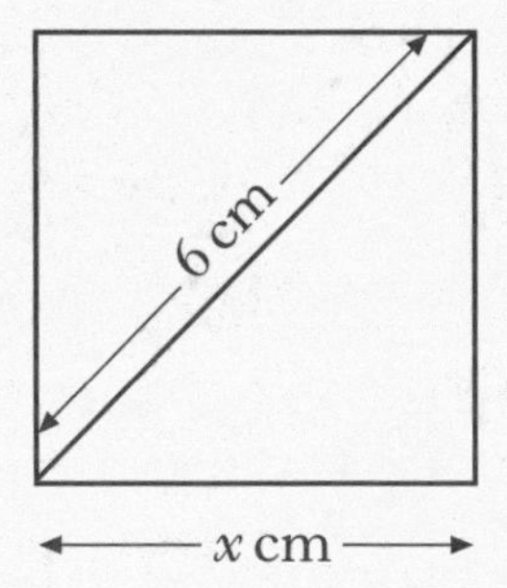

Calculate the value of x, giving your answer as a surd in its simplest form. 3

[Turn over

KU | RE

10. A relationship between T and L is given by the formula, $T = \frac{k}{L^3}$ where k is a constant.

When L is doubled, what is the effect on T? — RE 2

11. (*a*) A cinema has 300 seats which are either standard or deluxe.

Let x be the number of standard seats and y be the number of deluxe seats.

Write down an algebraic expression to illustrate this information. — KU 1

(*b*) A standard seat costs £4 and a deluxe seat costs £6.

When all the seats are sold the ticket sales are £1380.

Write down an algebraic expression to illustrate this information. — KU 2

(*c*) How many standard seats and how many deluxe seats are in the cinema? — RE 3

KU	RE

12. The diagram shows water lying in a length of roof guttering.

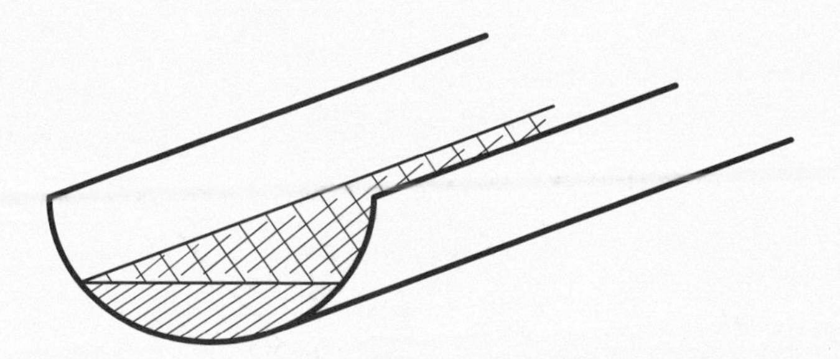

The cross-section of the guttering is a semi-circle with diameter 10 centimetres.

The water surface is 8 centimetres wide.

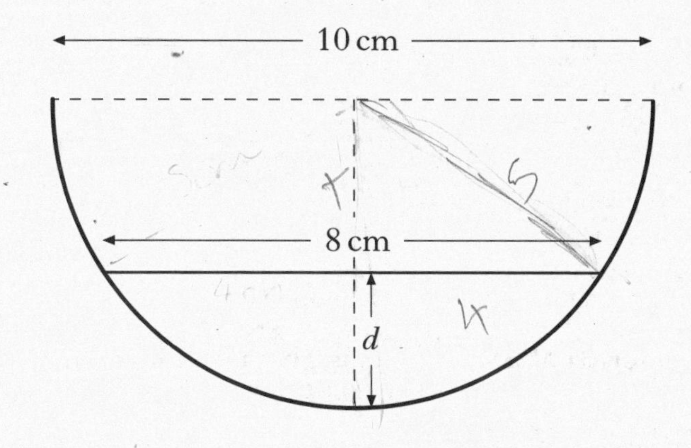

Calculate the depth, d, of water in the guttering. **4**

[Turn over for Questions 13 and 14 on *Page eight*

KU	RE

13. Part of the graph of $y = \cos bx° + c$ is shown below.

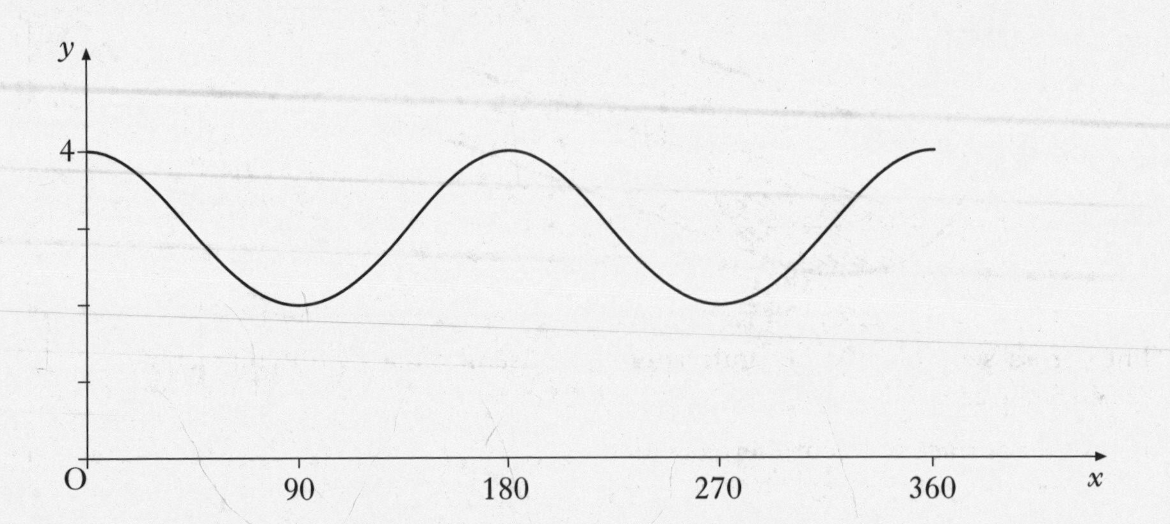

Write down the values of b and c. — KU 2

14. The **sum** S_n of the first n terms of a sequence, is given by the formula

$$S_n = 3^n - 1.$$

(*a*) Find the **sum** of the first 2 terms. — RE 1

(*b*) When $S_n = 80$, calculate the value of n. — RE 2

[*END OF QUESTION PAPER*]

C

2500/406

NATIONAL QUALIFICATIONS 2007

THURSDAY, 3 MAY
2.45 PM – 4.05 PM

MATHEMATICS
STANDARD GRADE
Credit Level
Paper 2

1 **You may use a calculator**.

2 Answer as many questions as you can.

3 Full credit will be given only where the solution contains appropriate working.

4 Square-ruled paper is provided.

LI 2500/406 6/31570

FORMULAE LIST

The roots of $ax^2 + bx + c = 0$ are $x = \dfrac{-b \pm \sqrt{(b^2 - 4ac)}}{2a}$

Sine rule: $\dfrac{a}{\text{Sin A}} = \dfrac{b}{\text{Sin B}} = \dfrac{c}{\text{Sin C}}$

Cosine rule: $a^2 = b^2 + c^2 - 2bc \cos \text{A}$ or $\cos \text{A} = \dfrac{b^2 + c^2 - a^2}{2bc}$

Area of a triangle: Area $= \frac{1}{2} ab \sin \text{C}$

Standard deviation: $s = \sqrt{\dfrac{\sum (x - \bar{x})^2}{n-1}} = \sqrt{\dfrac{\sum x^2 - (\sum x)^2/n}{n-1}}$, where n is the sample size.

	KU	RE

1. Alistair buys an antique chair for £600.

It is expected to increase in value at the rate of 4·5% each year.

How much is it expected to be worth in 3 years? **3** (KU)

2. Solve the equation

$$3x^2 - 2x - 10 = 0.$$

Give your answer **correct to 2 significant figures**. **4** (KU)

3. (*a*) During his lunch hour, Luke records the number of birds that visit his bird-table.

The numbers recorded last week were:

28 32 14 19 18 26 31.

Find the mean and standard deviation for this data. **4** (KU)

(*b*) Over the same period, Luke's friend, Erin also recorded the number of birds visiting her bird-table.

Erin's recordings have a mean of 25 and a standard deviation of 5.

Make **two** valid comparisons between the friends' recordings. **2** (RE)

4. Solve the inequality

$$\frac{x}{4} - \frac{1}{2} < 5.$$

2 (KU)

[Turn over

KU RE

5. Mark takes some friends out for a meal.

The restaurant adds a 10% service charge to the price of the meal.

The **total** bill is £148·50.

What was the price of the meal? **3**

6. Brunton is 30 kilometres due North of Appleton.

From Appleton, the bearing of Carlton is 065°.

From Brunton, the bearing of Carlton is 153°.

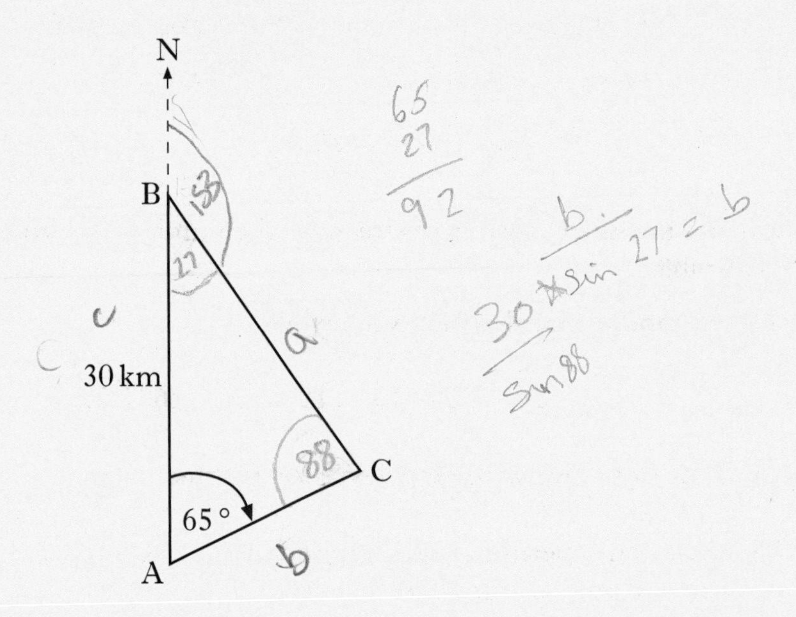

Calculate the distance between Brunton and Carlton. **4**

	KU	RE

7. A fan has four identical plastic blades.

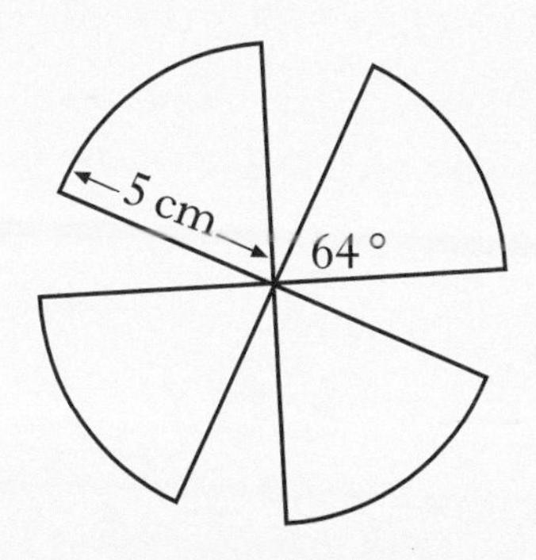

Each blade is a sector of a circle of radius 5 centimetres.

The angle at the centre of each sector is 64 °.

Calculate the **total** area of plastic required to make the blades. **3** (KU)

8. In triangle PQR:

- QR = 6 centimetres
- angle PQR = 30 °
- area of triangle PQR = 15 square centimetres.

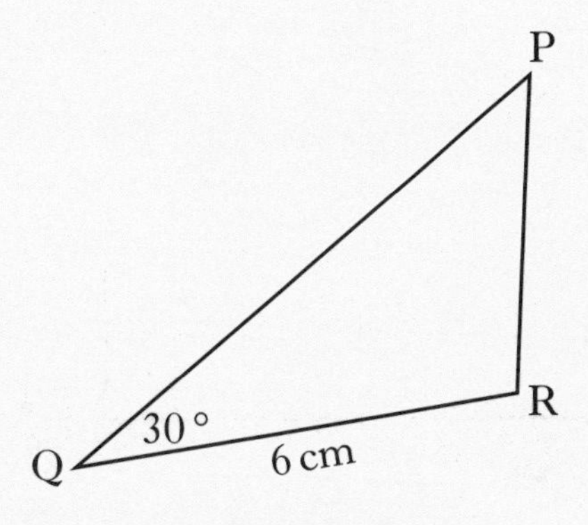

Calculate the length of PQ. **3** (RE)

[Turn over

KU	RE

9. To make "14 carat" gold, copper and pure gold are mixed in the ratio 5:7.

A jeweller has 160 grams of copper and 245 grams of pure gold.

What is the maximum weight of "14 carat" gold that the jeweller can make? **3** (RE)

10. Solve **algebraically** the equation

$$5\cos x° + 4 = 0, \qquad 0 \leq x < 360.$$

3 (KU)

11. (*a*) A decorator's logo is rectangular and measures 10 centimetres by 6 centimetres.

It consists of three rectangles: one red, one yellow and one blue.

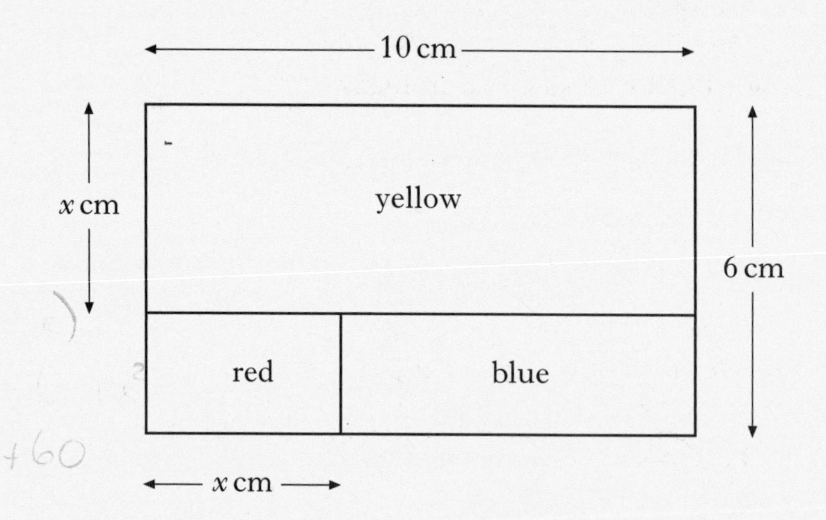

The yellow rectangle measures 10 centimetres by x centimetres.

The width of the red rectangle is x centimetres.

Show that the area, A, of the blue rectangle is given by the expression

$$A = x^2 - 16x + 60.$$

2 (RE)

(*b*) The area of the blue rectangle is equal to $\frac{1}{5}$ of the total area of the logo.

Calculate the value of x. **4** (RE)

KU	RE

12. (*a*) A cylindrical paperweight of radius 3 centimetres and height 4 centimetres is filled with sand.

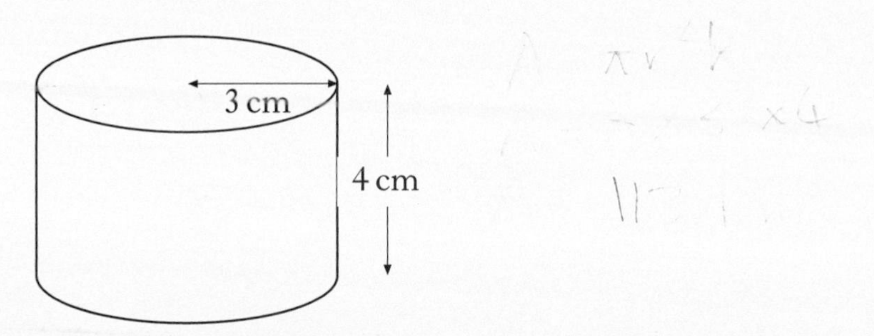

Calculate the volume of sand in the paperweight. **2**

(*b*) Another paperweight, in the shape of a hemisphere, is filled with sand.

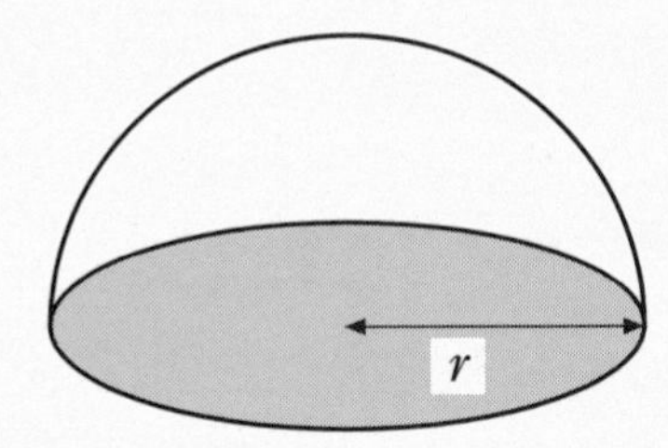

It contains the same volume of sand as the first paperweight.

Calculate the radius of the hemisphere.

[The volume of a hemisphere with radius r is given by the formula, $V = \frac{2}{3}\pi r^3$.] **3**

[Turn over for Question 13 on *Page eight*

KU	RE

13. The profit made by a publishing company of a magazine is calculated by the formula

$$y = 4x(140 - x),$$

where y is the profit (in pounds) and x is the selling price (in pence) of the magazine.

The graph below represents the profit y against the selling price x.

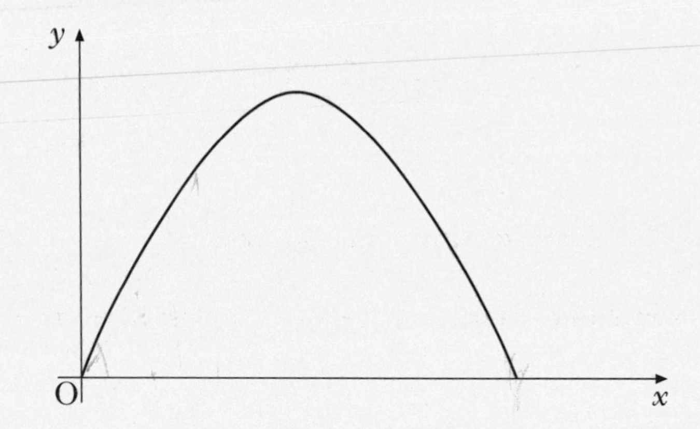

Find the maximum profit the company can make from the sale of the magazine. **4**

[*END OF QUESTION PAPER*]

[BLANK PAGE]

C

2500/405

NATIONAL QUALIFICATIONS 2008

THURSDAY, 8 MAY
1.30 PM – 2.25 PM

MATHEMATICS STANDARD GRADE

Credit Level
Paper 1
(Non-calculator)

1 **You may NOT use a calculator**.

2 Answer as many questions as you can.

3 Full credit will be given only where the solution contains appropriate working.

4 Square-ruled paper is provided.

LI 2500/405 6/31070

FORMULAE LIST

The roots of $ax^2 + bx + c = 0$ are $x = \dfrac{-b \pm \sqrt{(b^2 - 4ac)}}{2a}$

Sine rule: $\dfrac{a}{\sin \text{A}} = \dfrac{b}{\sin \text{B}} = \dfrac{c}{\sin \text{C}}$

Cosine rule: $a^2 = b^2 + c^2 - 2bc \cos \text{A}$ or $\cos \text{A} = \dfrac{b^2 + c^2 - a^2}{2bc}$

Area of a triangle: Area $= \frac{1}{2}ab \sin \text{C}$

Standard deviation: $s = \sqrt{\dfrac{\sum(x - \bar{x})^2}{n-1}} = \sqrt{\dfrac{\sum x^2 - (\sum x)^2 / n}{n-1}}$, where n is the sample size.

	KU	RE

1. Evaluate

$$24{\cdot}7 - 0{\cdot}63 \times 30.$$

2

2. Factorise fully

$$5x^2 - 45.$$

2

3.

$$W = BH^2.$$

Change the subject of the formula to H.

2

4. A straight line cuts the x-axis at the point (9, 0) and the y-axis at the point (0, 18) as shown.

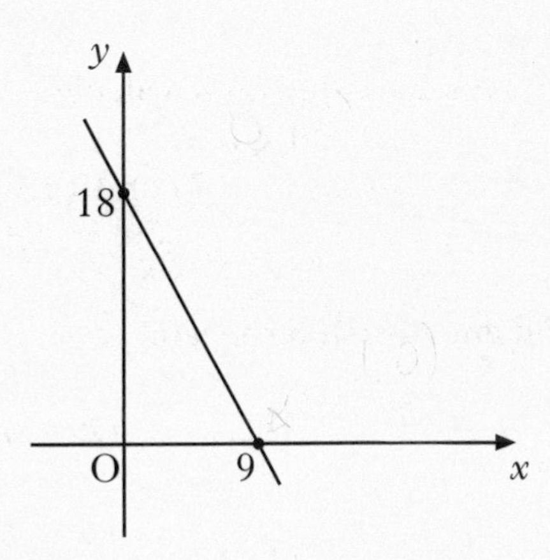

Find the equation of this line.

3

[Turn over

KU	RE

5. Express as a single fraction in its simplest form

$$\frac{1}{p}+\frac{2}{(p+5)}.$$

KU 2

6. Jane enters a two-part race.

(*a*) She cycles for 2 hours at a speed of $(x + 8)$ kilometres per hour.

Write down an expression in x for the distance cycled. KU 1

(*b*) She then runs for 30 minutes at a speed of x kilometres per hour.

Write down an expression in x for the distance run. KU 1

(*c*) The **total** distance of the race is 46 kilometres.

Calculate Jane's **running** speed. RE 3

7. The 4th term of each number pattern below is the **mean** of the previous three terms.

(*a*) When the first three terms are 1, 6, and 8, calculate the 4th term. KU 1

(*b*) When the first three terms are x, $(x + 7)$ and $(x + 11)$, calculate the 4th term. RE 1

(*c*) When the first, second and fourth terms are

$$-2x, \quad (x+5), \quad \_\_\_\_, \quad (2x+4),$$

calculate the 3rd term. RE 2

KU	RE

8. The curved part of the letter A in the *Artwork* logo is in the shape of a parabola.

The equation of this parabola is $y = (x - 8)(2 - x)$.

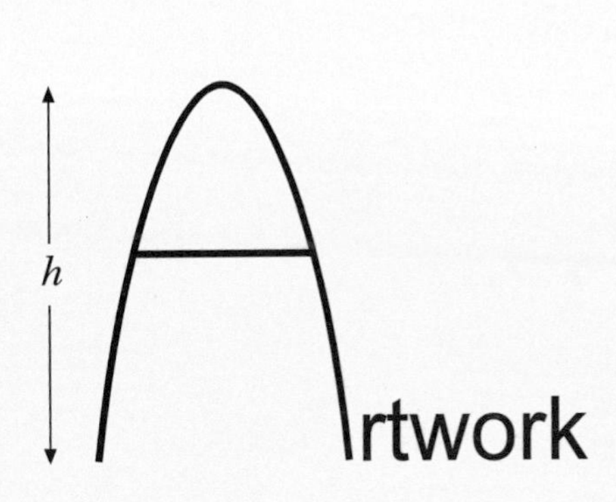

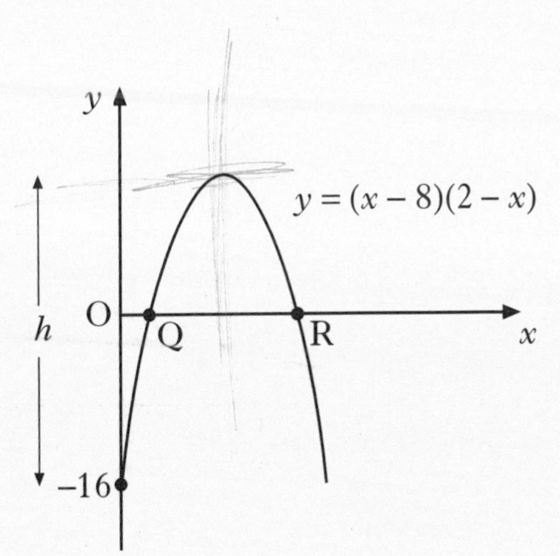

(*a*) Write down the coordinates of Q and R. **2** (KU)

(*b*) Calculate the height, h, of the letter A. **3** (RE)

9. Simplify

$$m^3 \times \sqrt{m}.$$

2 (KU)

[Turn over

KU	RE

10. Part of the graph of $y = a^x$, where $a > 0$, is shown below.

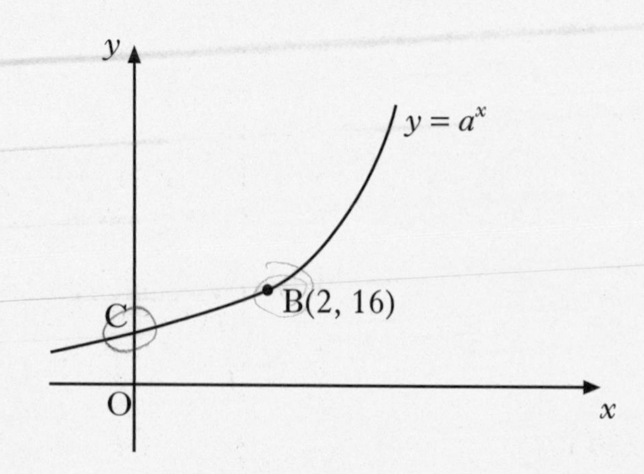

The graph cuts the y-axis at C.

(*a*) Write down the coordinates of C. **1**

B is the point (2, 16).

(*b*) Calculate the value of a. **2**

11. A right angled triangle has dimensions as shown.

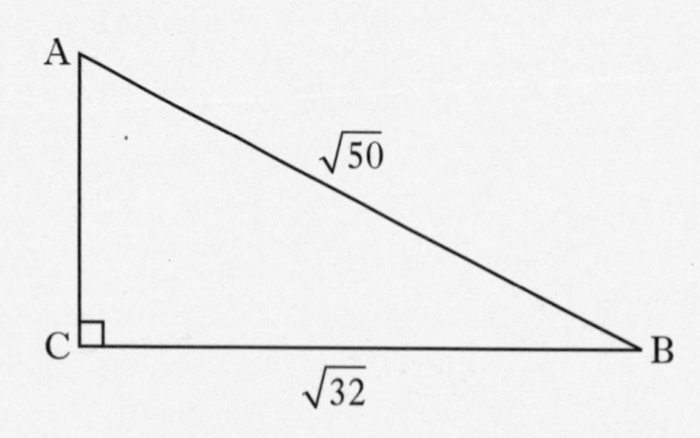

Calculate the length of AC, leaving your answer as a surd **in its simplest form**. **3**

KU	RE
3	
	3

12. Given that

$$x^2 - 10x + 18 = (x - a)^2 + b,$$

find the values of a and b. **3** (KU)

13. A new fraction is obtained by adding x to the numerator and denominator of the fraction $\frac{17}{24}$.

This new fraction is equivalent to $\frac{2}{3}$.

Calculate the value of x. **3** (RE)

[*END OF QUESTION PAPER*]

[BLANK PAGE]

C

2500/406

NATIONAL QUALIFICATIONS 2008

THURSDAY, 8 MAY
2.45 PM – 4.05 PM

MATHEMATICS STANDARD GRADE

Credit Level
Paper 2

1 **You may use a calculator**.

2 Answer as many questions as you can.

3 Full credit will be given only where the solution contains appropriate working.

4 Square-ruled paper is provided.

LI 2500/406 6/31070

FORMULAE LIST

The roots of $ax^2 + bx + c = 0$ are $x = \dfrac{-b \pm \sqrt{(b^2 - 4ac)}}{2a}$

Sine rule: $\dfrac{a}{\sin A} = \dfrac{b}{\sin B} = \dfrac{c}{\sin C}$

Cosine rule: $a^2 = b^2 + c^2 - 2bc \cos A$ or $\cos A = \dfrac{b^2 + c^2 - a^2}{2bc}$

Area of a triangle: Area $= \frac{1}{2}ab \sin C$

Standard deviation: $s = \sqrt{\dfrac{\sum(x - \bar{x})^2}{n-1}} = \sqrt{\dfrac{\sum x^2 - (\sum x)^2/n}{n-1}}$, where n is the sample size.

KU	RE

1. A local council recycles 42 000 tonnes of waste a year.

   The council aims to increase the amount of waste recycled by 8% each year.

   How much waste does it expect to recycle in 3 years time?

   Give your answer **to three significant figures**. **4**

2. In a class, 30 pupils sat a test.

   The marks are illustrated by the stem and leaf diagram below.

Test Marks

Stem	Leaf
0	9
1	6 6 7 8
2	0 4 5 7 9 9 9
3	2 2 3 5 5 6 8
4	0 2 3 4 5 5 7 7 8
5	0 0

n = 30   1 | 6 = 16

   (*a*) Write down the median and the modal mark. **2**

   (*b*) Find the probability that a pupil selected at random scored **at least** 40 marks. **1**

3. In a sale, all cameras are reduced by 20%.

   A camera now costs £45.

   Calculate the **original** cost of the camera. **3**

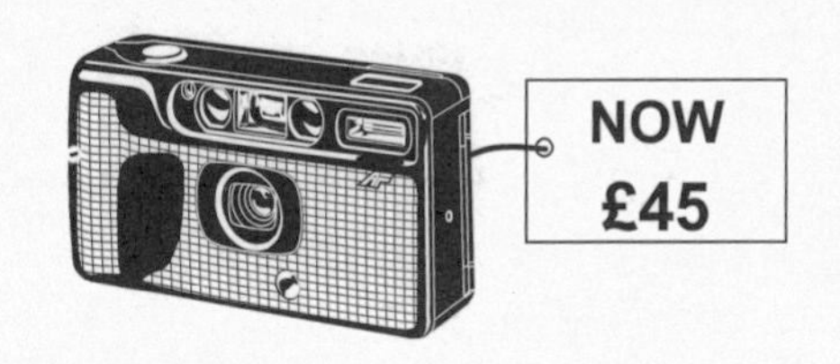

[Turn over

KU	RE

4. Aaron saves 50 pence and 20 pence coins in his piggy bank.

Let x be the number of 50 pence coins in his bank.

Let y be the number of 20 pence coins in his bank.

(*a*) There are 60 coins in his bank.

Write down an equation in x and y to illustrate this information. **1** (KU)

(*b*) The total value of the coins is £17·40.

Write down another equation in x and y to illustrate this information. **1** (KU)

(*c*) Hence find **algebraically** the number of 50 pence coins Aaron has in his piggy bank. **3** (RE)

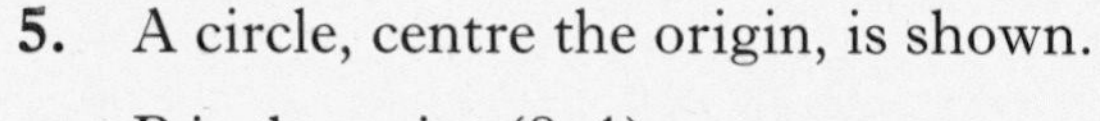

5. A circle, centre the origin, is shown.

P is the point (8, 1).

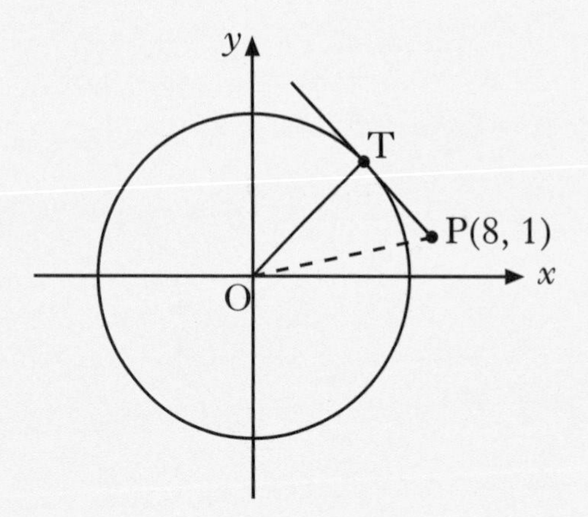

(*a*) Calculate the length of OP. **2** (RE)

The diagram also shows a tangent from P which touches the circle at T.

The radius of the circle is 5 units.

(*b*) Calculate the length of PT. **2** (RE)

KU	RE

6. The distance, d kilometres, to the horizon, when viewed from a cliff top, varies directly as the square root of the height, h metres, of the cliff top above sea level.

From a cliff top 16 metres above sea level, the distance to the horizon is 14 kilometres.

A boat is 20 kilometres from a cliff whose top is 40 metres above sea level.

Is the boat beyond the horizon?

Justify your answer. 5

7. A telegraph pole is 6·2 metres high.

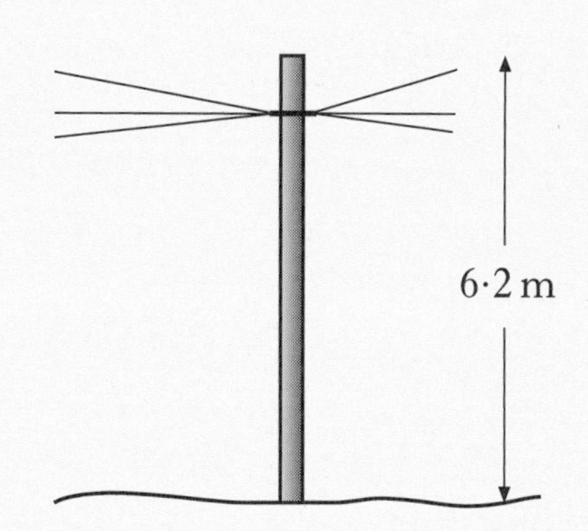

The wind blows the pole over into the position as shown below.

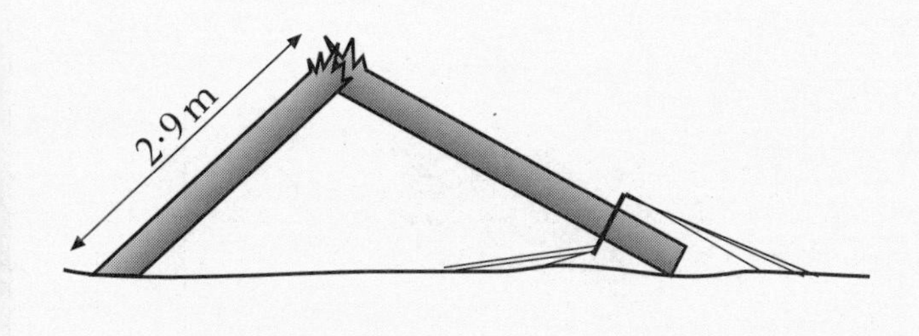

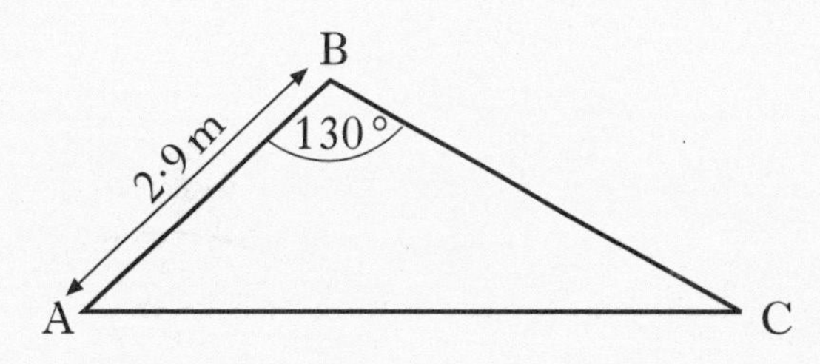

AB is 2·9 metres and angle ABC is 130 °.

Calculate the length of AC. 4

[Turn over

	KU	RE

8. A farmer builds a sheep-pen using two lengths of fencing and a wall.

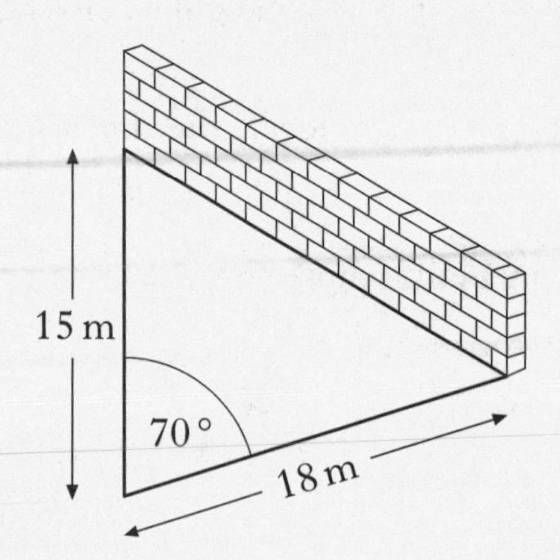

The two lengths of fencing are 15 metres and 18 metres long.

(*a*) Calculate the area of the sheep-pen, when the angle between the fencing is 70 °. **3** (KU)

(*b*) What angle between the fencing would give the farmer the largest possible area? **1** (RE)

9. Contestants in a quiz have 25 seconds to answer a question.

This time is indicated on the clock.

The tip of the clock hand moves through the arc AB as shown.

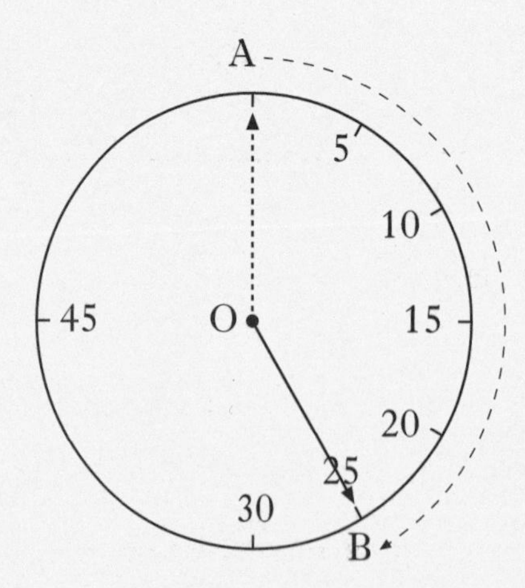

(*a*) Calculate the size of angle AOB. **1** (KU)

(*b*) The length of arc AB is 120 centimetres.

Calculate the length of the clock hand. **4** (RE)

KU | RE

10. To hire a car costs £25 per day plus a mileage charge.

The first 200 miles are free with each additional mile charged at 12 pence.

CAR HIRE

£25 per day

- first 200 miles free
- each additional mile only 12p

(*a*) Calculate the cost of hiring a car for 4 days when the mileage is 640 miles. **1** (KU)

(*b*) A car is hired for d days and the mileage is m miles where $m > 200$.

Write down a formula for the cost £C of hiring the car. **3** (RE)

11. The minimum number of roads joining 4 towns to each other is 6 as shown.

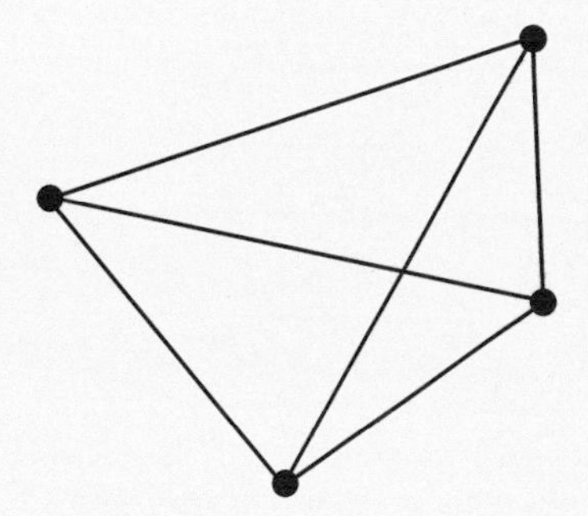

The minimum number of roads, r, joining n towns to each other is given by the formula $r = \frac{1}{2}n(n-1)$.

(*a*) State the minimum number of roads needed to join 7 towns to each other. **1** (KU)

(*b*) When $r = 55$, show that $n^2 - n - 110 = 0$. **2** (RE)

(*c*) Hence find **algebraically** the value of n. **3** (RE)

[Turn over for Question 12 on *Page eight*

KU	RE

12. The diagram shows part of the graph of $y = \tan x°$.

The line $y = 5$ is drawn and intersects the graph of $y = \tan x°$ at P and Q.

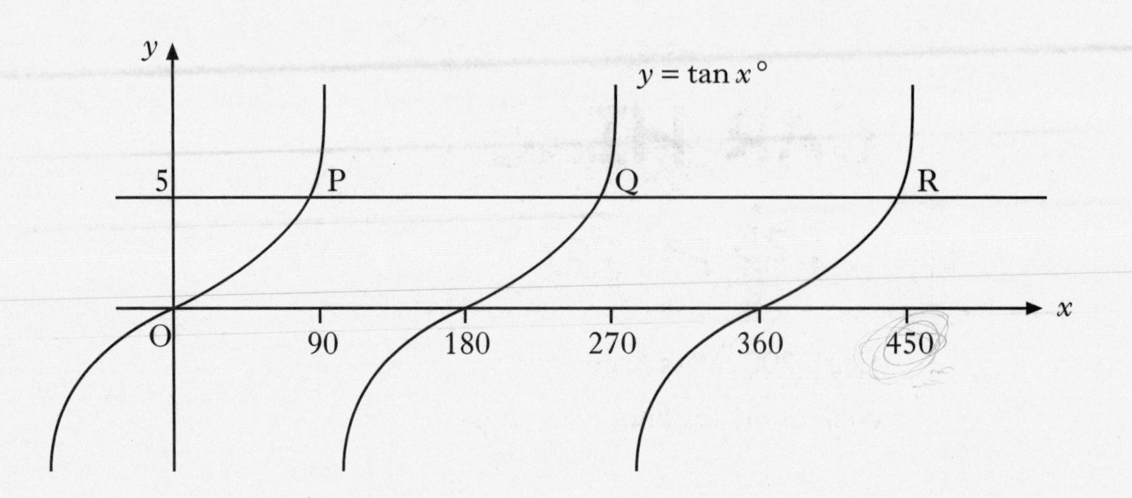

(*a*) Find the x-coordinates of P and Q. **3**

(*b*) Write down the x-coordinate of the point R, where the line $y = 5$ next intersects the graph of $y = \tan x°$. **1**

[*END OF QUESTION PAPER*]